# The Old English and Modern Game Fowl

*by P. Proud*

**with an introduction by Jackson Chambers**

*This work contains material that was originally published in 1903.*

*This publication is within the Public Domain.*

*This edition is reprinted for educational purposes and in accordance with all applicable Federal Laws.*

*Introduction Copyright 2018 by Jackson Chambers*

# Self Reliance Books

Get more historic titles on animal and stock breeding, gardening and old fashioned skills by visiting us at:

# http://selfreliancebooks.blogspot.com/

# Disclaimer

This book was written in an age when cock-fighting was widely acceptable throughout society. In many places throughout the world, cock-fighting has been made illegal.

The material presented herein is intended to be strictly for educational purposes with the purpose of enlightening Game Fowl breeders about the history of their breed. Publication of the material is neither an endorsement, nor a criticism of its contents. This book is presented as part of large series of educational material on the history and raising of numerous chicken breeds for utility or exhibition purposes.

As the reader, please consider it your duty to become familiar with local, state, provincial and federal laws relating to the subject matter contained herein before attempting to utilize any of the information presented.

As the author, publisher and retailer cannot control how the reader utilizes the historical information presented in the pages herein, they hereby disclaim any liability to any party for any loss, damage, disruption or other liability that may be incurred by the reader's misuse of this material.

## *Introduction*

I am pleased to present this third title in the "Game Fowl" series.

The work is in the Public Domain and is re-printed here in accordance with Federal Laws.

Though this work is a century old it contains much information on poultry that is still pertinent today.

As with all reprinted books of this age that are intended to perfectly reproduce the original edition, considerable pains and effort had to be undertaken to correct fading and sometimes outright damage to existing proofs of this title. At times, this task is quite monumental, requiring an almost total "rebuilding" of some pages from digital proofs of multiple copies. Despite this, imperfections still sometimes exist in the final proof and may detract from the visual appearance of the text.

I hope you enjoy reading this book as much as I enjoyed making it available to readers again.

Jackson Chambers

## PREFACE.

THE requests which have reached me for the publication in a more permanent form of the articles contributed by Mr. Proud to *The Feathered World* have prompted this reprint of papers which have thus evidently been proved useful. The chapters have been carefully revised, and, with the addition of Illustrations and two Coloured Plates, they should certainly be of real service to the novice, and, perchance, convey a hint or two to even older fanciers.

The inclusion within the same cover of an exact reprint of that old and standard work on Game Fowl, *The Cocker*, was not arrived at without serious consideration. To bowdlerise the book by cutting out all that pertained to a sport which the laws of a more humane age forbid, would have been to rob the work of much of its quaintness, and of a good deal of information. Not, therefore, as an incentive to cock-fighting, but as an aid to the breeding of healthy, vigorous stock have I reprinted the pages of that worthy old fancier, William Sketchley, Gent., as he is so pleasantly described

## PREFACE.

on his 1814 title-page. As a history of practices prevailing a century since, *The Cocker*, with its curious frontispiece of an old-time cock-fight, may well contrast with Mr. Proud's story of its modern successor, and the illustration of a game contest at Birmingham in 1902.

THE EDITOR,
*The Feathered World.*

# CONTENTS.

## PART I.

### OLD ENGLISH GAME.

| | PAGE |
|---|---|
| Black-Reds | 3 |
| Spangles | 6 |
| Duckwings | 7 |
| Blue-Reds | 8 |
| Piles | 11 |
| Brown-Reds | 12 |
| Blacks and Whites | 13 |
| Brassy Wings | 13 |
| Birchens | 13 |
| Muffs and Hennies | 13 |
| Standard of Perfection | 14 |
| Breeding and Management | 18 |
| Perches | 21 |
| Feeding | 21 |
| Dubbing | 22 |
| Preparing for Exhibition | 25 |

## PART II.

### MODERN GAME.

| | |
|---|---|
| Chief Characteristics | 28 |
| Black-Reds | 31 |
| Brown-Reds | 34 |
| Piles | 37 |
| Duckwings | 43 |
| Birchens | 46 |
| Whites and Blacks | 46 |
| Standard of Perfection | 48 |
| Hatching and Rearing | 53 |
| Feeding | 55 |
| Diseases | 56 |

## ILLUSTRATIONS.

|  |  | PAGE |
|---|---|---|
| *Coloured Plate*—Old English Game | . . . . | *Frontispiece* |
| Black-Reds „ | . . . . . | 9 |
| Wheaten Pullet „ | . . . . . | 17 |
| Black-Red Cock „ | . . . . . | 19 |
| Spangled Cock „ | . . . . . | 23 |
| *Coloured Plate*—Modern Game | . . . . . | *face* 27 |
| Brown-Red Pullet „ | . . . . . | 41 |
| Varieties of „ | . . . . . | *face* 48 |
| A Game Contest in Modern Times | . . . . . | 61 |
| A Game Contest in Olden Times . | . . . . . | 64 |

# PART I.

## OLD ENGLISH GAME.

WHEN, where, and how the Old English Game fowl originated is now probably beyond the power of any man accurately to decide. Some would have us believe that it is a descendant of the Jungle Fowl of India or Ceylon, but we have really no definite information on the point of its ancestry.

The ancient classical writers speak of fighting fowl in their day. It is probable that, for centuries, pugilistic birds entertained their owners with the display of their talents, and that gradually the birds were bred towards that size, strength and agility which best suited their work.

What is now called Old English Game was certainly never the development of a single lifetime. We incline to the opinion expressed above, that the breed was gradually worked age after age into the points found half a century ago, and the type altered generation after generation, advancing towards what was considered the ideal of perfection for fighting purposes. In this book it is not my intention to treat upon the pugilistic qualities of this noble breed, but to give my readers some practical and helpful knowledge of the Old English Game fowl as an exhibition bird; and at the same time try and establish an up-to-date standard of points required in the exhibition Old English Game fowl of the twentieth century.

That the breed has existed pretty much as we see it now for a very long time goes without saying, and we predict that, so long as fowls obtain any hold upon the fancier, Old English Game will have a strong position in the lists. It may not be that classes at shows will always be maintained, but, apart from pugilistic tendencies, when these latter are added in the appeal to the British fancier's taste, the breed is certain to find a *locus standi* in some hearts, and to be bred as it has been bred so long for the love of the thing, rather than for any honour and glory in the prize lists. At the present time, the Old English fowl is, however, bidding fair to drive the Modern Game into less room at our shows, and possibly almost to annihilate it. A perusal of the entries of any show of to-day (with the exception, perhaps, of Birmingham, the yearly rendezvous of Modern Game fanciers) cannot for one moment leave this matter in doubt. Seldom have secretaries of shows any reason to cancel the Old English classes; whereas it is a matter of common knowledge that at many shows the Modern Game classes have to be either cut down or cancelled to save a substantial loss to the societies. At some shows during the past season the Old English Game classes have been the flower of the exhibition, entries beyond precedent, and the quality superb. And it is not owing to the fickleness of fanciers that the Old English has so monopolised things. It is true that in the foremost ranks of its breeders to-day stand some of the most prominent Modern Game breeders, who only a year or two ago scoffed at the "mongrel" fowl making its appearance in the show pen, but these have been won in part over to the Old English fowl through sheer necessity. The Modern English Game has been so inbred for colour, and to secure fine round bone, that it can hardly be advanced further, and at the present time is in some danger of falling into neglect, through the great difficulty of rearing the very delicate chicks. In these busy times men have neither time nor money to throw away upon fowls whose chicks require so much capital, care and labour, and whose progeny after all may break down and leave the breeder *minus* a season's fruits. The Old English Game fowl has not been thus inbred for colour or fine bone. The colour is a secondary consideration only, and bone is in request. The breed therefore admits of frequent crossings of strains, and in consequence gives fuller results, which again give greater satisfaction.

The varieties of Old English Game are legion, and can easily be run up to a score or more, as, for instance, black-reds, ginger-reds, spangles, duckwings, brassywings, blue-reds, blues, piles, duns, greys, Furnesses, brown-reds, birchens, creels, cuckoos, greys, whites, blacks, muffs and several others.

Four of these, however, are more commonly met with than others, and so will be more fully dealt with here than the others. They are black-reds, spangles, duckwings and blue-reds.

## Black-Reds.

The Old English black-red is identical in colour with the Modern black-red. The face, head, comb, wattles and ears should be a bright, healthy cherry-red, with no touch of white in the latter. The head must be powerful, fairly short, beak strong and slightly curved, eyes rich ruby-red, a yellow, white or pearl eye being a disqualification, although at some of our big shows birds possessing white and yellow eyes have been placed in the prize-list by so-called specialist judges; but this is decidedly wrong, as a light eye denotes weakness of sight, and, apart from this, nothing looks better than a rich, ruby-red eye in a Game fowl. A bird with a bad eye should under no circumstances be bred from, however good he may be in other respects.

The neck hackle should be as bright an orange as possible; back, wing bow, and saddle, rich bright crimson, with saddle hackle light orange to match the neck; wing bars and wing butts, steel blue-black; secondaries, a rich chestnut to end of feather; breast and thighs, black; tail, green-black, although some strains have more or less white in tail, which is allowable in Old English, and only counts against the bird when it comes to a deciding point, and then only to a very small extent.

We now come to the leg colour, which to-day gives rise to dispute as to what is the correct leg colour of each variety. In black-reds I should certainly favour white legs, also in spangles and blues. Although I should certainly not discard a good bird of the two former varieties with yellow legs, still if there were two birds of almost equal merits in the class, and the one had white legs and the other yellow, I should give my vote to the former, and I think that such a course would find favour with the majority of Old English breeders. The shin-bone should be as round and smooth as possible, not flat in front, as sometimes seen, and which denotes weakness of limb. Flat shins are a serious defect, and should count against the bird in the show pen. Toes should be long, firm and straight, the back toe placed directly opposite front middle toe, the point resting firmly on the ground, with no tendency to come to the side, which is termed duck-footed, and as such is liable to be disqualified. A Game fowl with deformed feet is useless, either as an exhibition bird or for stock. The bird should be low on leg, powerful in thigh, broad and deep in chest, not like those narrow hollow-chested specimens that do duty in the show pen at some of our shows, and which are simply a bundle of feathers with a beautiful colour, as unlike a Game fowl as it is possible to imagine. Not only broad in shoulder, but broad and short in back, limbs set on firmly and well apart, giving the bird a good footing to enable him to hold his own against any foe. The wings should be short and well rounded, not flat-sided, which is very objectionable, tail full and fairly long, with drooping sickles and side hangers. Such a bird would stand a grand chance for a champion prize in the show pen. The hen to match this cock may be either wheaten or partridge. Both, to my mind, are beautiful birds. The wheaten is so called because the general top colour resembles that of red wheat. It is a soft reddish cinnamon grey, whilst the breast, thighs and underparts are a pale fawn. In face, comb, wattles and ears she should be like her lord and master, a bright healthy red. The neck hackle should be a dark golden, or even inclined to chestnut, and the tail should be well clipped together, and not carried too high. The shoulders should be prominent, not hidden amongst the feathers of her breast, as we sometimes see, breast broad and full, breast-bone perfectly straight; a twisted or dented breast-bone is a disqualification both in cocks and hens. Back perfectly flat, and the body brought to a wedge shape. Legs short, thighs muscular, shins round, smooth and white.

The partridge hen resembles her in everything save colour. In this particular she requires to be a soft, even light shade of drab, inclined to the yellower rather than the chocolate side; in fact, any ruddy feathers about her wings, as on the bow and bar, would somewhat spoil her chances in the show pen, if birds with clear drab-feathered wings were opposed to her. Moreover, she must be nice and smooth in her pencilling. By pencilling I do not mean such as you see on partridge Cochins. If a feather of a good-coloured black-red Game hen be very closely examined you will see some very fine, minute black markings running irregularly across the feather. They are extremely fine, and therefore on a good feather hardly perceptible. If they show as really distinct marks or blotches it may be taken that they are too coarse. When clear elsewhere, an otherwise good bird will be spoilt by coarseness on top feathers, which run down over the secondaries of the wings. This is the first place one looks to to form an idea as to the quality of the colouring of a show hen, for quite 50 per cent. fail here, although perfect in colour in all other parts. Her feet and legs, like the wheaten, should be white. Partridge hens are very difficult to breed free from rustiness, or shaftiness. What I mean by shaftiness is when the white shaft in the centre of the feather shows up too white and distinct. This is accounted for by the introduction of wheaten blood to brighten the colour of the cocks.

For breeding black-reds I prefer two pens, the one for cockerel and the other for pullet breeding. And, as will be seen by and by, the cockerel pen can be made more or less to do duty for breeding other than black-red cockerels where space is limited. To breed black-red cocks select the brightest-coloured male bird you can, with sound black breast and good bays on wings. To him match typical wheaten hens, white-legged like the cock, and like him good and sound in all the typical points of shoulders, breast, thighs, back, head, eye, beak, etc., which belong to a perfect specimen. Let your first object be perfect shape, broad at shoulder, short in back, low and cobby in build. Never breed from long-backed, flat-sided specimens, or the outcome of your season's labour will be a perfect failure. See that the legs are strong and firm, set well apart, head powerful, brilliant red eye, and that full daring appearance which a good Old English Game fowl has, especially when viewed from the front. From such a mating good black-red cockerels should result, as perfect in colour as the best Modern Game of the present day, and typical in other points.

If a good-bodied blue hen be introduced into this pen then good blue-breasted red cockerels should result, and there is no prettier or more taking variety than the blue-red cocks. Not only can blue-reds be produced in addition to black-reds from this pen, but by introducing a good-shaped, clear-hackled duckwing hen you can look forward to having two or three really good golden duckwing cockerels as well, and all from the same pen. Thus you see that cockerels of three distinct varieties can be bred from the same sire, and at the same time. This is a great advantage where space is limited, which is often the case with the working-man fancier, who has only perhaps a small back garden at his disposal. Not only three varieties of cockerels, but also a very fair percentage of wheaten pullets, would result also from this pen, with probably a useful blue pullet or two.

In breeding for partridge pullets, a darker, more even, and more brick-colour-red throughout kind of cock is required. The hens should be as perfect in colour as it is possible to secure, although really sound-coloured partridge hens are very rare indeed. It is all the better if it be known beforehand that the cock is descended from perfect-coloured hens of a pure pullet strain. This is half the battle in pullet breeding. See that the hens are clear on wing from rust or shaftiness, and as fine and soft in pencilling as possible. Next to sound colour in hens, see that they approach near to the ideal *in shape;* better far sacrifice a little in point of colour, but never by any means overlook shape in hens when mating up your breeding pen, no matter what variety it is, for on the female side shape is very important. Although from such a pen you would breed good sound-coloured pullets, it would be found that the cockerels will come a little dark for the show pen. Still, with first-class judges, if these cockerels excel in character and type, they will stand a very good chance, for colour, as I have said before, is but a secondary consideration. As a rule these pullet-breeding cockerels will be found more shapely and harder in feather than the bright-coloured ones.

It is no use sacrificing the true points of an Old English to mere colour. It is false to the breed. This has been done from time to time, I know, by the introduction of brown Leghorn blood into a strain, but the effect has been not only to brighten the colour, but also to introduce a featheriness and softness inimical to the characteristics of this grand old breed. Hard feather, hard condition, good shape, style and carriage should not thus be sacrificed to a mere whim for colour. Though I am no sympathiser with cockfighting, at the same time, if we are perpetuating the true characteristics of a fighting breed, a breed whose very existence was due to the desire for hard sport on the part of its admirers, it is evident that all points which, theoretically, in this day seem to have been from necessity a *sine qua non* in the old pit days, should take prominence over mere beauty points, otherwise our fowl is no longer typical of the true Old English Game fowl. Therefore I say once more, avoid the reduction of the breed down to dubbed brown Leghorns, and the breeder will be safer with white-legged birds rather than yellow. There can be no Leghorn blood in the former, but possibly there may be a lot in the latter, and for this reason I should give preference to white-legged birds when it came to a deciding point in the show pen.

### Spangles.

Of all the varieties they are, in my estimation, the most beautiful. Of course, in shape, build, and general characteristics the cock should be identical with the black-red cock, and have a good sound ruby or blood-coloured eye. This is somewhat a difficult point to get in spangles, as they are apt to favour the yellow or daw eye, both of which are serious defects. This can easily be remedied by breeding only from rich ruby-eyed stock birds on both sides, and discarding those with a tendency to yellow or light. The hackle, back, saddle, saddle hackle and wing bow should be the same ground-colour as that of a good black-red, but should be freely and as evenly as possible spangled with white and black, which gives a very handsome appearance. Add to this a similar breast and thighs to the black-red, but as evenly spangled with white as possible. The chestnut secondaries of the black-red are here replaced by pure white, and the tail and sickles white, whilst the side hangers of the tail are black. Many good tails, however, have the feathers mostly white, with some few inches of the sickles and hangers a rich green glossy black. The wing bar is white with an occasional black feather; but it is better if both bars and secondaries are quite white.

The hen to match the cock may be either a wheaten spangle or a partridge spangle. The latter are the handsomer, as the contrast of drab, black, and white feathers is the more striking. The spangling of black and white should be as even as possible, and can be so bred by using evenly-spangled birds in the breeding pen. Should, however, there be a tendency on the part of the chicks to come too light-coloured, this may be remedied by using a dark-coloured cock and evenly-spangled hens of the medium shade. The one great advantage in breeding spangles is that both sexes can be well produced from the same pen with a facility that does not exist in any other variety. The leg colour of the spangle may be either white or yellow, but, all other points being equal, preference should be given to white-legged, but only as a deciding point where two birds are of equal merit.

### Duckwings

come next in order. The cock should be silvery white or creamy white in neck hackle, some of the oldest breeders preferring the latter, free from dark striping down these feathers; his wing bow, back and saddle, a rich brassy maroon colour, having two or three rich shades of tints, and giving the impression of extra brightness and richness. The legs are white, tail and breast a good green-black, with steel-blue wing bar, and clear white on secondaries. This is often spoilt by unsightly chocolate markings, a defect which is found in quite 50 per cent. of the present-day Game, both Modern and Old English. The eye should be red, as in the black-red, although a general failing in this variety is to be dark-eyed; still, though a defect, it is not a disqualification, and should not count too heavily against the bird in the show pen. In size, shape, carriage, etc., he should follow the black-red ideal. The hen is a lovely steel-grey in top colour and top feathers of tail, the softer the tint the better. She should be free from any blotchiness or coarse markings generally, and especially on the wings; her breast should be a pale salmon-red grey, hardly so deep as the black-red hen, as this is exceedingly difficult to obtain; still, the deeper the better, so long as the body colour, wings and back are a nice soft steel-grey. As a rule, when the deep salmon breast is obtained the body colour is found to be too dark and hard in colour. Legs should be white, and white only; her eye, a rich ruby red; and hackle, a silvery or creamy white, finely striped with black. She is a very taking bird when seen to perfection.

To breed the best duckwing cockerels use preferably a sound-coloured black-red cock and pale-hackled silver duckwing hens. See that the cock is especially sound in his breast, bars, and black generally, with good sound white wing ends, the bay running right out to the wing ends. By insisting on this point you will ensure a perfect white wing end in your cockerels, a thing much to be desired. The following year I should reserve the best-coloured cockerel, and mate him back to the old hens, preferring hens with sound breast colour. Good-coloured cockerels can be produced equally as well this way as by the black-red.

Duckwing pullets are bred from sound-coloured duckwing hens and a rather pale-coloured duckwing cock, the top colour favouring the yellow or straw colour rather than the rich maroon, and preference should be given to a bird that has been bred from a sound-coloured hen. The cockerels from this pen would be mostly silvers, the best of which should be retained for future pullet breeding. Some of the best-coloured duckwing pullets have been bred from a pullet-bred duckwing cock already described, and mating him to pure soft-coloured black-red or partridge hens from a pullet strain. A very good percentage of sound-coloured pullets can be obtained in this way.

## BLUE-REDS,

or what are sometimes termed blue-piles and blue-duns, now demand a few words. The cock is a strikingly handsome bird, with his golden top colour and his slate-blue where the black-red is black and the pile white, with blue and white tail,

MR. FRED ANDERTON'S BLACK-RED OLD ENGLISH GAME.

white legs and toes. The hen is self-coloured, sound slate-blue throughout.

The breeding for colour in this variety is yet somewhat unsatisfactory, as one hardly knows what colour one may expect from special matings. Very often pure colours are bred together and with good results— *i.e.*, pure blues on both sides; but a good sound-coloured blue-breasted cock and wheaten hens have been known to produce excellent blue-breasted cockerels. The best pullets probably are bred from blue on both sides, though occasionally a decent one will come from the blue-breasted cock and wheaten cross. Another successful mating might be found for cocks by using a sound-breasted black-red cock with blue hens; this would ensure good breast colour in the blue cockerels. Still it is a bit of a lottery producing tip-top specimens of this variety, as more fixity of colour character is yet required in this variety. As time goes on this will be acquired, and then we shall expect to see blue-breasted reds taking a foremost place at our exhibitions.

## Piles,

although a very popular variety with Modern Game fanciers, find few admirers in Old English circles, hence it is very rarely we meet them in the show pen, except perhaps at our first-class shows, and then they are few and far between. Why this is so we cannot say, for when produced on the right lines it is quite as handsome as some of the more popular varieties.

Piles can be bred either from piles on both sides or by mating a pale-coloured or white hen to a black-red cock. As the black-red has made such good progress towards perfection, I should be inclined to adopt the latter course to ensure good results, as the type and character would thus be more greatly enhanced, than by breeding from piles which at the present day are far behind the black-red in Old English Game characteristics. In using the black-red cock it is important to notice that he is sound in his breast, free from lacing, and a rich black, the same on shoulder points and wing bars. The bays or wing ends should also be a sound chestnut, carried out right to the end of the wing. This is absolutely necessary to ensure a rich deep bay on the pile cockerel, a striking contrast to the white wing bar. The pile cock should be white where the black-red is black, in other respects he is the same colour. Legs and feet white or yellow, former preferred; eyes red. The pile hen should be a creamy white all over except hackle and breast; former should be a light lemon; breast pale salmon, shading lighter towards the thighs. Some prefer the deep salmon-coloured breast, but such invariably leads to rosiness on the wings, which we do not care to see in exhibition pile hens. Still, a little rosy colouring on the wing of an otherwise good Old English pile hen would not count very strongly against her in the show pen.

Piles are the best colour as a rule first season, especially the cockerels. The second and subsequent years they become marble-breasted, and are therefore unsightly in the show pen.

Nothing looks prettier, to my idea, than a rich top-coloured pile cock, with a deep chestnut bay unbroken in colour, and a spotlessly white breast, the two latter points being the most difficult to obtain, the contrast being so great. Still, by careful breeding and judicious mating for these desired points, a good percentage can be obtained.

### Brown-Reds

are more popular to-day than piles, especially so at our Northern shows, whilst in type and character they are rapidly approaching the black-reds. At the Birmingham Show the cup for the best off-coloured Old English hen was secured by a brown-red.

The colour of the Old English brown-red cock differs so far greatly from the colour of the Modern brown-red: in the former the top colour is a deep orange approaching to red, whilst the latter is pale lemon. The eye of the brown-red also differs from all others, being a very dark brown, almost black, whilst the legs and beak are a dark willow, the darker the better, face dark red, inclined to mulberry, breast evenly laced with brown. The hen should be a green-black in body colour, head and neck hackle coppery, striped with black, breast laced as in the cock. Face, eyes and legs dark.

Brown-reds breed true to colour, and good birds of both sexes can be bred from the same pen, providing that the cock selected is not too dark in his top colour, and the hens free from lacing on back and shoulders, as the bright colour in the Old English brown-red cock is not nearly so important as in the modern variety, where two breeding pens must be used in order to obtain anything like satisfactory results.

### Blacks and Whites

are seldom met with, except at our big events: the former are decidedly the more popular of the two. The cock should be a glossy black all over, free from any other colour whatever. Face, dark or red; eyes, ditto; legs, black. The hen should match the cock in every respect.

The whites do not appear to find favour with Old English exhibitors, probably on account of their soft feather, and are very rarely met with. The plumage should be white; eyes and face, red; legs, white or yellow.

### Brassywings

would appear to be descended from blacks, as the only perceptible difference is a little brassy or dark-lemon colour on the back and shoulders of the cock.

### Birchens

are offshoots from the duckwings and brown-reds, and should be very similar to the latter, with the exception that where the brown-red is a deep orange the birchen should be a silvery white; the breast lacing should be the same. In all other respects he should resemble the brown-red. Birchens may be bred by crossing a silver duckwing cockerel and a brown-red hen, and this cross is also beneficial in improving the colour of the brown-red cockerels; but the pullets from such a cross are invariably faulty in top colour, being shafty and laced on top and wing. Therefore, great care should be exercised that this cross does not get mixed in your pullet-breeding strain, or the result will be disastrous.

Besides many of the off-coloured varieties, which are far too numerous, and, as a rule, are only the sports of other established varieties, there are

### Muffs and Hennies.

The former are found in nearly every colour of the more popular varieties. They derive their name from the thick growth of feathers under the throat in both male and female. The best of these I have met with in my travels are blues and black-reds. They are rather taking in the show pen, and frequently find favour with many of the present-day judges.

To the young fancier taking up Old English, I would recommend him to start with one of the first three or four varieties if he desires to be successful in the show pen, for it is only at the very largest exhibitions that classes are provided for the off-coloured varieties, and nothing definite has yet been arrived at in reference to the proper standard of these outside breeds. To those with only a limited space at their disposal black-reds and spangles offer the best opening, and these two popular varieties are the most successful in the show pen, and the easiest to breed true to colour and type.

The following is the Standard of Perfection, reproduced by kind permission of The Poultry Club:—

### General Characteristics of Cock.

*Head and Neck.*—Head, medium length and tapering; beak, strong at base and slightly curved; eyes, large, bright and prominent, full of expression and alike in colour; comb, single, small, evenly serrated, erect and of fine texture; face, fine texture to match the comb and wattles; ear lobes, to match the comb and wattles as nearly as possible; wattles, fine texture and small; neck, long and very strong at junction with body; neck hackle, wiry long feathers, covering shoulders.

*Body.*—Breast, broad and well developed, indicative of constitutional vigour, straight breast-bone; back, short, broad across the shoulders, and flat, tapering to the tail; belly, small and compact; wings, long, full and round, inclining to meet under the tail, amply protecting the thighs and furnished with very hard quills.

*Tail.*—Sickle feathers abundant, broad, curved main feathers with hard strong quills.

*Legs and Feet.*—Thighs, short, thick and muscular, well set and held wide apart; shanks, medium length, finely and evenly scaled, not flat on shins; toes, four on each foot, should be clean, even, long and spreading, the back toe standing well backward and flat on the ground; spurs, low on the leg.

*General Shape and Carriage.*—Bold and smart, the movements quick and graceful, proud and sprightly as if ready for any emergency.

*Handling.*—Clever, flesh firm but corky and light, mellow and warm with strong contraction of the wings and legs.

*Size and Weight.*—5 lb. to 6 lb.

*Plumage.*—Hard, glossy and firm.

### General Characteristics of Hen.

*Head, Neck, Body.*—As in the cock.
*Tail.*—Inclined to fan shape and carried well up.
*Legs and Feet.*—As in the cock.
*General Shape and Carriage, Handling, Plumage.*—As in the cock.
*Size and Weight.*—4 lb. to 5 lb.

### Colour in Black-Breasted Red Game.

*In both Sexes.*—Beak, in character with legs; eyes, red;* face, bright red; legs, any sound self colour.

*In the Cock.*—Neck hackle and saddle, orange red, free from dark feathers; back and shoulder coverts, deep red; wing bow, deep red; wing bar, rich dark blue; secondaries, bay colour; primaries and wing ends, black; breast and under parts, black; tail, black with lustrous green gloss.

*In the Hen* (Partridge).—Neck, golden red streaked with black; back and wings, partridge colour; breast and thighs, shaded salmon colour; tail, black shaded with brown.

### Colour in Bright or Ginger-Red Game.

*In both Sexes.*—Beak, in character with legs; eyes, red;* face, bright red; legs, any sound self colour.

*In the Cock.*—Neck hackle and saddle, light golden red, free from streaks; back and shoulders, bright red; wing bow, bright red; wing bar, rich dark blue; secondaries, bay colour; primaries and wing ends, black; breast and under parts, black shaded with brown; tail, black or black shaded with brown.

*In the Hen.*—Neck hackle, golden red; back and wings, a darker shade of wheaten than the breast; breast and thighs, light wheaten; tail, black with a shading of brown.

### Colour in Brown-Red Game.

*In both Sexes.*—Beak, dark horn; eyes, dark; face, red or dark; legs, dark.

*In the Cock.*—Neck and saddle, orange-red streaked with black; back and shoulders, dark red; wing, dark brown or

---

* In white-legged birds, daw eyes and a few white feathers in wings and tail are quite allowable and in character; the hackle should also be white at the roots of the feathers next the skin.

black; breast and thighs, brown or brown marked and shaded with black; tail, black.

*In the Hen.*—Neck hackle, black striped or shaded golden; body, black or of a uniform brown mottle; tail, black.

### Colour in Red Pile Game.

*In both Sexes.*—Beak, in character with legs; eyes, bright red;* face, brilliant red; legs, white, yellow or willow.

*In the Cock.*—Neck and saddle, orange or chestnut red; back and shoulders, deep red; wing bar, white; secondaries, bay on the outer edge of feathers and white on the inner edge and tips, the bay colour alone showing when wing is closed; primaries, white; breast and under parts, white; tail, white.

*In the Hen.*—Neck, light chestnut; breast and thighs, chestnut, shading lighter towards thighs; rest of body, white.

### Colour in Silver Duckwing Game.

*In both Sexes.*—Beak, in character with legs; eyes, red;* face, red; legs, yellow, white, olive or blue.

*In the Cock.*—Neck and saddle, silver white, free from dark streaks; back and shoulders, silver white; wing bow, silver white; wing bar, steel blue; secondaries, white on outer web, black on the inner web and tip of feathers, the white only showing when the wing is closed; primaries, black; breast and thighs, black; tail, black.

*In the Hen.*—Neck, silver striped with black; back and wings, dark grey; breast and thighs, pale fawn; tail, grey and black.

### Colour in White Game.

*In both Sexes.*—Beak, yellow; eyes, red or pearl; face scarlet red; plumage, white throughout; legs, white or yellow.

### Colour in Black Game.

*In both Sexes.*—Beak, dark; eyes, red or dark; face, red or dark; plumage, glossy black throughout; legs, sound self colour.

### Colour in Brassywings.

*In both Sexes.*—Same as in the Black Game, with the exception of a little dark lemon on shoulders of cock.

* See note on eyes and legs, p. 15.

## POINTS FOR JUDGING. 17

### Colour in Spangled Game.

*In both Sexes.*—Beak, in character with legs; eyes, red or daw; face, bright red; plumage, either black, red, blue or buff spangled with white, the spangling as even as possible; tail, black and white; legs, self colour or mottle.

Value of points in Old English Game:—

| Defects. | Deduct up to |
|---|---|
| Defects in head, 4; beak, 4; eyes, 6 | 14 |
| ,, neck, 6; back, 8 | 14 |
| ,, breast and body | 12 |
| ,, wings | 6 |
| ,, thighs, 4; shanks, 6; spurs, 2; feet, 9 | 21 |
| ,, plumage | 7 |
| ,, carriage | 10 |
| ,, colour | 8 |
| ,, handling | 8 |
| A perfect bird to count | 100 |

Serious defects for which a bird should be passed:—Crooked or humped back, crooked breastbone, wry tail, flat shins, duck feet, bad carriage, rotten plumage, or any unsoundness.

OLD ENGLISH GAME WHEATEN PULLET.
1st, Haltwhistle; 2nd, Hallbankgate, etc.

(Mrs. Mordaunt Lawson's.)

## BREEDING AND MANAGEMENT.

IN the breeding of Old English Game a few general principles may be laid down.

There must, in the first place, be nothing approaching to feebleness, undersize, or disease in the parent birds. They must be the most vigorous of the vigorous. Any degeneracy on this head will inevitably undo the breeder. The birds selected should handle hard and light, and have a springy, elastic, corky feel about them. The breast-bone must be straight, the breast full, the feet sound, head full of fire, with a ruby eye, thighs short, round and stout, and back wedge-shaped. The age should not as a rule exceed three years for the cock, and especially in his third season should his harem be kept few in number. At any time, the fewer the hens or pullets, the more vigorous are the chickens likely to be. Cockerels should never be less than nine months, and, where size, bone and strength be required, should be over twelve months old, and a cross should always be resorted to as soon as it is seen that the yard is losing ground on these points. A safe rule is to mate a two-year-old cock with pullets March or April hatched, and a cockerel not under nine months with two-year-old hens. The in-breeding necessary to retain type and colour should never exceed three years at the most, or the strain will become enfeebled.

Old English thrive naturally where there is an unlimited grass run. They are especially suited to the northern counties of England, and some of the best specimens have hailed from Cumberland, Westmoreland, Lancashire, and the borders of Northumberland. The northern climate produces harder feather than the more relaxing conditions of the south country. By roosting the birds in trees at night, greater vigour and better health with hard feather is secured; and such sleeping places, even in the depth of winter, secure for the birds great immunity from roup and other ailments, whilst condition and bloom are everything that can be desired. The one drawback to such a system is the necessity for withdrawing birds from these roosting places for exhibition purposes. They then have to pass a night or two in a warm room, and it would be disastrous to turn the birds out again into the open air to sleep after their return.

MR. HARRISON WEIR'S OLD ENGLISH GAME COCK.

## Perches.

Proper perches are an absolute necessity to all kinds of Game fowl. Branches of trees being round and covered with bark, and generally selected by the bird of a size that is easily and firmly grasped, afford a good illustration of what is required when perches have to be provided indoors.

For bark, wrappings of hemp or other soft material may be substituted, so that the breast-bone of the fowl comes by no injury, as a crooked breast destroys all a bird's chances in the show pen, and is also an unsightly thing—difficult to carve too, on a table. Never perch the birds too high from the ground. Give them plenty of room to fly down, or the soft yielding bone of the breast of young birds may easily become deformed. The force with which the breast of a bird can come in contact with the hard floor, when the flight is too steep, is surprising, and from this cause many a good bird has been utterly ruined. Not only the breast-bone but also the feet have been completely crippled by being perched too high. How often do we notice in the show pen birds with bumble feet, enlarged joints, etc., thrown out of the prize lists completely, and all this caused simply by perching the birds too high and allowing them to fly down on to the hard floor of the roosting house. For Game fowl the perches should never exceed three feet from the ground, and then the floor should have a liberal covering of peat moss litter or chaff, to soften the fall. Never use flat perches, or your birds will become duck-footed, and as such are useless as exhibition specimens. Therefore it will at once be seen that proper perching of all Game fowl is highly important, and should nowise be overlooked.

## Feeding.

To keep the birds in good condition feed them well and regularly. In winter time the food should be of a warmth-giving nature, and as a first feed in the morning nothing is better than biscuit meal, well scalded with boiling water, and afterwards, when properly soaked, mixed with good Scotch oatmeal and fine sharps until of a crumbly nature. Good, sound English wheat at night cannot be beaten. See that it is sound and dry, for this is most important. Nothing is more injurious to fowls than damp, unsound corn with any tendency to mildew; if there be any suspicion of dampness, put it to the oven to dry and harden; the harder the corn the better.

For a change from wheat—and change of food is always beneficial—give short stout oats or barley. Barley is best given during the coldest weather, and wheat in the summer months. Don't throw the food down indiscriminately in a heap, but scatter it about judiciously, and only give enough, so that every grain is picked up. As soon as the birds give over eating rapidly don't throw more down; then by the time they are satisfied all the food will have been picked up. Food allowed to remain on the ground for long causes disease, and upsets the birds' liver and digestion. Therefore it is much better to rather stint the birds than give them too much.

### Dubbing.

The dubbing of Game fowl is a practice which has obtained from earliest time, and I think must and will be continued as long as there are Game fowl to dub. Game are very pugnacious by nature, and therefore dubbing is justified by necessity, for it would be impossible to keep any number of cockerels together after puberty is reached without resorting to the practice. Moreover, the irritation caused by the constant pecking and tearing of the comb, during a succession of fights, is infinitely more painful than that produced by the needful operation of dubbing, which can easily be completed in two or three minutes at the most.

In dubbing Old English it is not necessary to cut so near to the head as in dubbing Modern Game, for by leaving a slight margin of the base of the comb, it gives the head a stronger appearance than when cut low down. Curved scissors should always be used, which are manufactured for the purpose, and cost about four or five shillings. The bird should be held by an assistant whilst the operator first cuts off the lobes and wattles. The comb is the last to come off, and should be taken with one long sweeping cut. In this way very little pain is given. A sponge with a little warm water is then applied to the head, and the wounded parts may be smeared with vaseline the day following the operation. In from eight to ten days all traces of the operation will have departed, beyond just a little paleness here and there, where the more gristly parts of the comb and wattles have been.

**MR. HARRY S. HASSALL'S SPANGLED OLD ENGLISH GAME COCK.**

## PREPARING FOR EXHIBITION.

This is a very important part of the programme. Many a good bird has been spoiled through want of training. A bird that dashes about the pen when approached stands a very small chance of success in the show pen. The judges of to-day have so much work to perform when judging in the time allotted to them that they cannot stand on one side and wait until the bird has finally settled down; they have no alternative but to pass the bird with a remark in their judging book that he is "wild in pen". I have met scores of them when judging—birds that had they been properly trained would have stood very high in the prize list.

In the first place, procure one or two 27-inch wire show pens. Any poultry appliance maker will supply them for about half-a-crown. Put the bird in at night time and allow him to get used to the pen before beginning operations. Let him remain perfectly quiet. The next morning get a little bread and milk and approach the pen slowly, speaking kindly to the bird at the same time. Then feed him through the wires very gently. In a little while he will come boldly up to eat the bread and milk. The same night repeat the same process, and after he has partaken of the food from your hand, gently open the door and stroke the bird slowly down the breast from the throat. Do this a few times, and you will be surprised how soon he learns to stand perfectly still and appears to enjoy your caresses. Then quietly and gently stroke him on the back from the hackle to the tail, talking softly to him all the time, and at the same time giving him little tit-bits of bread-sop or lean meat. In three or four days the bird will have become quite docile and will stand boldly up to the front of the pen when you approach without the slightest fear. After getting him settled down to his work, use a small cane instead of the hand. A week will be found quite long enough for even the wildest bird, and once properly trained they never forget it, no matter if they don't see the inside of a pen for months. In getting the bird into proper exhibition form the following loaf should be used, a few pieces steeped in sherry, and given to the bird last thing at night. Procure a half-pennyworth of balm, one pennyworth of sugar candy, ¼ lb. lump sugar, five whites of eggs, two yolks, a piece of fresh butter the size of a walnut, and 1½ lb. of flour. Crush the sugar candy and loaf sugar well together. When baked properly this loaf will keep sweet and good for two or three months if kept wrapped up in a cloth, and will be found one of the finest pick-me-ups for turning out a Game fowl as fit as a fiddle, as the saying goes. To the young fancier whose birds have hitherto been beaten in condition this is a valuable tip.

Before despatching the bird to the show, it is necessary to sponge the head and face with warm water. The legs also should be well soaked in hot water as hot as you can comfortably bear the back of your hand in. Let them soak for three or four minutes, then take a piece of soap and a nail brush and thoroughly scrub the legs and feet till all traces of dirt have disappeared. There will always be a little dirt under the scales, but this can easily be removed by getting a piece of wood and sharpening it to a fine point, then insert under the scales from one side to the other gently, so as not to break off the scales. This will remove every particle of dirt from under the scales. Dry the legs, then polish well with a piece of rough dry flannel, rubbing briskly until they are as smooth as glass. This process will put an excellent polish on any leg, whether white or yellow, and where the latter is inclined to be pale, it will materially improve the colour.

And now I think I have told my readers all that is worth knowing about this grand old breed as an exhibition bird. The Old English Game is a noble fowl, strong, fearless and ever on the alert, beautiful in plumage, graceful in carriage, and a champion amongst other fowls. The hen is a model mother, both as regards hatching and looking after her chicks, and bold to a degree when danger threatens her little ones. She is a fair layer of richly-flavoured eggs, and both sexes afford the finest quality of poultry flesh to be found upon the table. Whether we look at the Old English Game from the exhibition side or the utility one, there is everything to give us satisfaction. The chicks are easy to rear, which again is in its favour; in fact, it matters not in what way we view this fine old breed, it stands out of the pack of all other varieties as the Ace of Trumps.

# PART II.

## MODERN GAME.

### INTRODUCTION.

IN presenting my readers with the following Chapters on Modern Game, their points and manner of breeding, I do so in the sincere hope that I may be able to infuse a little more enthusiasm into this portion of the Fancy than seems at present to exist.

We naturally ask, "How is it that the Modern Game has dwindled down to its present dimensions?" for of all classes at present-day shows the Modern Game are nearly always the most disappointing.

Take, for instance, Birkenhead Show, September, 1902, with a Game Club judge wielding the wand. The result was humiliating. An average of about 2 per class. The reason is soon explained. It is the old tale over again, and, wherever it has to be told the same results follow. *Modern Game have fallen into too few hands.* It is a fatal mistake. What encouragement is there nowadays to take up Modern Game?

The novice, that truly necessary individual to the success of any breed, finds no incentive to action. He does not know the ins and outs of breeding like the old hands. Very often his cash-box is not so deep that he cannot find the bottom. But when he has found the bottom or thereabout in his endeavour to make a start in the Modern Game Fancy, the superior knowledge of the old hands, coupled with the fact that they *never* let the best out of their keeping, even if the "till" were emptied a dozen times over, brings about the result that at the end of a few seasons' breeding, the young aspirant still finds himself nowhere.

The deck-sweepers are upon him. He cannot stand, with all his British pluck, being constantly knocked down before he has had time to rise to his knees as it were. The British fancier has always looked to the cash side, as well as the Fancy side, and he naturally expects some adequate return for capital, perseverance, time, trouble, expense of rent, labour, houses and food. Now if anything that I can here say will tend to the amelioration of this sad state of things, I unflinchingly say it. Let present breeders consider that they cannot eat their cake themselves and have it. They have a right to win on their experience and superior knowledge of breeding, but if they are men of heart they will see that this should be sufficient without reserving to themselves the major portion of the handicap, *viz.*, the *superlative* breeding stock as well.

Now I am satisfied that, with greater liberality in the way I have here indicated, we should soon have a different order of things. I dare not predict it in other breeds which have gone down as the result of deck-sweeping, but Game is different.

In this work I have used my best efforts to put the tyro on a footing with the old hands by going carefully into the mating, breeding and preparing for exhibition. From what I have written he should easily be able to recognise the type required, and if he be allowed the privilege, for an equivalent of the present currency, by such birds being placed within his reach, all will go well, and we shall once more see the Modern Game forging to the front, and the classes at our shows monopolising that extended area of pens we were wont to see at our best shows some few years ago.

CHIEF CHARACTERISTICS.

Proceeding to the individual varieties as they appear at present-day exhibitions, the principal ones are five in number, viz., black-reds, brown-reds, piles, duckwings and birchens. I will take them in the order here given, after a statement of the chief characteristics common to all and every variety.

Beginning with the cock, his general appearance to a novice would be that of an extremely long-legged bird, finely built, with a well-carved away body, an apology of a tail, and any amount of length of neck and head, but wherewithal scanty, hard and narrow of feather all over. Descending to particulars, we find this general type exactly carried out in the best specimens.

The head should be long and narrow, snaky, as it is termed, with a strong, long, well-curved beak. Very little skull should appear above the eye, the less the better. The eye itself should be full, bold and daring, giving a nonchalant

## CHIEF CHARACTERISTICS.

expression to the clear red face and throat, which should be free from all coarse feathers.

The neck sets off the head. It should be long and fine, with the narrowest and shortest feather possible, so short that at the base of the neck it fails to reach and cover the shoulders at all. Whatever its colour may be, the saddle feathers should be short and narrow. The breast should be broad and carried erect, and the long thighs so placed that when the bird puts itself to its fullest reach, neck, breast and legs are not far out of a perpendicular straight line. And this power of reach is a *sine qua non*. A bird that carries itself duck fashion is utterly worthless. What is wanted is the capability, as the housemaid recently put it in one of Mr. Bernard Partridge's admirable sketches, to stretch. When applying for a situation she asked her mistress, "Shall I have to hand the things round at the table, *or do you stretch?*" A bird that cannot stretch up and far is useless.

Then the feet are most important items. The toes must be long and straight and firmly placed upon the ground. The hind toe sometimes fails to touch the ground, at other times it is not planted down in a straight line opposite in direction to the centre middle toe, but takes a forward position inwards towards the first toe. Such defects amount to absolutely disqualification points, and the cook is, strictly speaking, the only person to deal adequately with such a specimen. Now, the shanks also require a word. They should be fine and, above all things, round. No genuine fancier or judge will tolerate flat shins; it is a sign of hereditary weakness. Mark well the distance from the back of the thighs to the end of body; it should be short and well cut away. When a good specimen is in hand it is wonderful how little there is to feel. If the bird be short here, the wings will be short, the back short (as it should be), and the shape, in most cases, correct.

These short-backed ones are generally very broad at the shoulders, with shoulder butts standing out prominently, giving a strong but keen, agile impression ready for anything.

The general appearance of the body is more or less that of a box iron for laundry purposes, save and except that from the square front to the point the body is cut away upwards, and after the manner of a wedge. The tail is another important feature. Its carriage must be low. The tail proper is short and narrow in feather, tightly held together, and by no manner of means formed after the fashion of a Hamburgh. The sickle feathers are short and as narrow as they can be got; the side feathers, or hangers as they are called, are a mere apology, narrow and short as they can be grown.

The tail is a feature that has for many years been greatly attended to, and breeders of twenty years ago, who rejoiced to think they were then possessed of fine whip tails, would simply find themselves nowhere in present-day competitions.

Now, I have endeavoured to give the general characteristics of an ideal cockerel. The pullet follows, so far as a hen can follow the cock, on the same lines. She must be fine in head, the neatest and smallest of combs, with long, fine neck, bright red face and ears, and the shortest of feathering upon the neck, as everywhere else. She, too, should have full, broad chest, round, long legs and thighs, toes as in the cockerel, and general shape of the body the same. The less lumber there is about her the better. She wants the same flat-iron shaped body, prominent wing butts, short wings, and a fine whip tail of very narrow feathers held well together. There is about her a general appearance of being extremely well groomed, owing to the tightness, shortness and narrowness of her feathering, and the upright, sprightly carriage which no other breed of fancy poultry possesses. If she is for the most part coloured on body and tail, as in the black-red and duckwing varieties, the well-groomed appearance is aided by extreme evenness of ground colour and pencilling, but this will be entered into more fully as I take each variety in turn.

I conclude this chapter by advising all would-be breeders to get their stock of the best. Wait until you can pay a decent price, then fix upon some prominent winning strain belonging to a thoroughly reliable man, and then go to him, explain your wants, and discreetly put yourself upon his honesty. Satisfy yourself whether you wish to breed cockerels or pullets. It is better that you know your own mind in this matter at the start, as the very best specimens, as a rule, do not come from a mixture of cockerel and pullet strains, but this again is a point that will be dealt with in treating separately of each variety. Then, when you have made a start with a certain strain, keep to that strain. Do not court disaster by chopping about from strain to strain, mixing up this and that for the mere reason that you think you have the opportunity to buy a good bird cheaply from some other strain. Blood will tell in the long run. It always has, it always will, so be advised, and, once embarked on a strain, keep to it as long as you can possibly rear healthy chickens, and when fresh blood is required go to some one who commenced with, and has kept, to the same strain as yourself. Never breed from an unhealthy bird, never use one in the breeding pen that has a flaw amounting to a disqualification. There may be faults, there will be. There never was a bird of any kind absolutely correct, but never perpetuate a fault by double breeding, *i.e.*, never have the same fault in both cock and hen. Give your breeding stock the best value of food obtainable. Do not coop them up more than is absolutely necessary, and if they have to be limited see that such treatment does not run them into fat and render abortive your attempts to breed successfully from them. Keep them in fair condition, but flat-iron birds mean few eggs and delicate chickens. It is simply marvellous how Game birds confined can, in a very short space of time, become internally one mass of fat, overlaying all the internal organs, and sometimes resulting in sudden death. At breeding time they are generally voracious in appetite, and it requires much determination on the part of owners of some strains to run them on somewhat short commons, that disaster may not ensue.

### Black-Reds.

I now take up the variety known as black-reds, as it has now for so long held the premier place at our exhibitions, and, in almost all works treating of English Game, takes precedence over other varieties. The exhibition cock should follow the general lines I have laid down for all varieties of Game as regards shape, size, style, reach, etc.—that is to say, head long, fine and narrow, with eye alert, placed well up in his head, so as to come as near to the top of the head as possible, with long, strong, horn-coloured beak, giving the head, roughly speaking, more of the general contour of that of the Common Pheasant than the ordinary barn yard fowl. The unfeathered skin of the face must be a bright cherry red, smooth, and of good texture, and the whole impression derived from the head must be what is called "snaky," a conglomeration of boldness, cunning and alertness. The neck should be long and fine, hackle very short, just touching the shoulders, of the brightest orange possible, corresponding in some measure with the hackle feathers of the saddle, which, however, as a rule, are a lighter and brighter orange still.

The front view of a good bird is especially captivating. The breast is black, free from any blots, spangles, or lacings of rusty colour, and so carried that neck, breast and legs come into an almost perpendicular line. The rich black should extend down to the shanks, but in many cases the hocks betray more than a suspicion of rusty feathers. Sometimes it is a case of "where they are," often a naked place to show "where they were".

The shoulder points should be noticeable as the bird shows itself from the front, standing out square and prominent, and the thighs should be fully visible up to and including the joint with the body, in fact, as the bird reaches to its full height the place of the joint is clearly discernible, and gives the impression of extra tallness. Many a bird never gets itself so that the thigh can be seen in this manner, but the upper part seems hidden in the body. This arises from deficiency of limb, or want of reach, or both; but such a bird is severely handicapped in the show pen nowadays.

Now, if we look at the bird sideways, we want no long boat shape nor straight long-sided wings, no rust on bars or shoulder points. He must be short from neck to tail, shoulder points jet black, with brightest crimson on shoulders and across back, not peppered or broken with black, but standing as whole and entire as possible in one unbroken patch, whilst the bay of the wing should be a bright chestnut, going through to the end of the wing, and accompanied by a lustrous steel black bar and steppings. The tail should be carried fairly low, with the feathering as narrow and short as possible, a rich green black and well whipped up together. Turn him round now and view him from behind. The back is flat-iron shaped, the more so the better. Coming round to the front again we examine shanks, toes, nails, etc. The shanks and toes should be as long as possible, willow or olive, and the back toe well placed on the ground. If he is duck-footed, wring his neck. He can have but one use, viz., the table. The nails should be long and strong.

Now, if you possess a black-red up to this description, show him by all means. He will give a good account of himself.

The exhibition hen is in many respects like her lord. She must be very tall, very reachy, broad in front, high and prominent in shoulders, flat-iron shaped in body, fine and short in feather, good and sound in shanks (no flat shins, they are an abomination), long toes, shanks and thighs, the latter set well apart, not knock-kneed, a very common fault. She is, of course, snaky in head, like the cock, but more refined, but quality and colour of the eye must be the same. The hackle, extremely short, should be a nice pale golden colour, with a black stripe down the centre of the feather (a narrow line on each side of the shaft), and dark caps should be avoided.

The breast may be described as a broken salmon red, which should descend from throat downwards, getting paler in colour towards the thighs. Some pullets are much paler in breast colour than others, but the deeper and richer salmon is most admired. The top colour is somewhat difficult to describe without a coloured plate, and still I have never yet seen a really good representation of the colour of a black-red Game hen. Generally it is spoken of as partridge, by which would be understood a brown finely pencilled with black—so finely that the pencilling is almost lost to view. Yet there is a particular hue and cast of colour about it more of a drab than a brown. The pencilling referred to above is a great point. There must be no dark bars, blotches, or lacing about it anywhere, and the place to look most particularly is the end of the secondaries or upper feathers of the wings, which are the last in the adult stage to take the place of the chicken feathers. It is a very purely coloured bird where these can be found smooth and soft, free from streaks, lines or blotches.

Again, the wing must be of the same soft tone of colour as the back, and not rusty or flushed more or less with a deeper tone of breast colour. Such would be fatal, if strongly found, to any exhibition bird. Colour counts so much that no bird with any great flaw therein would have much chance nowadays in a show pen, although valuable as a cock breeder.

We now come to mating. Two pens are better than one.

For cock breeding colour can be got very bright in two ways. One by the employment of wheaten hens, and the other by using the very palest clear capped and clear hackled partridge hens. These are shades lighter (and often more yellow drab colour) than exhibition pullets. Many breeders swear by the latter, and such would be my choice in mating up the cock-breeding pen.

Make certain that the pullets are bred from a cock breeding strain, or light partridge blood, or you will find yourself with a lot of dull coloured cockerels on your hands by-and-bye, utterly unfit for exhibition.

Game breeding stands no happy-go-lucky work of any kind. However, it is possible to breed some winners of both sexes from one pen by using two cock-breeding hens or pullets, very pale hackles and red-sided, and the number of sound coloured exhibition hens, as pale in hackle as possible. Then, if the cock is fairly bright of a pullet strain, you can reasonably expect a few winners of both cockerels and pullets from this pen. Still, this course should only be pursued when space is limited, and I would prefer to confine my attention to either cockerel or pullet breeding, and not to both. Where both are pursued the greatest care has to be exercised, and the breeding stock kept well apart. Should a wheaten-bred cock cross into the pullet strain, it would cause such havoc as no one, in these rapid days of advancement, could afford, either in time or cost, to eliminate. It would be a case of a fresh start, for the two could not be induced to amalgamate with anything like satisfactory results for many years.

In mating up your cock-breeding pens, see that you have shape, style and reach in your pullets and bright colour and reach in the cock. Never breed from short-legged, faulty-shaped pullets, or your labour will be in vain. In pullet breeding the hens and pullets must be perfectly sound in colour, free from rustiness or shaftiness, and not too dark. They must excel in shape and be as tall as possible, whilst the cock should be a brickish red colour, one even shade from neck to tail, not shaded off in hackles, as in the cock breeder. Follow these lines and you won't get far wrong. Black-reds, if properly mated, are easily bred, and produce a larger percentage of winners, perhaps, than any variety of Game.

### Brown-Reds.

Brown-reds will now engage our attention. This is a most handsome variety, and come next to black-reds in popularity. For many years brown-reds have held their own in the show pen, and have, like the black-reds, commanded fabulous prices. Mentioning high prices brings to our mind the sale of Mr. Hugo Ainscough's Birmingham winning black-red cock to the king of Game fanciers, viz., Capt. Heaton, for £200 some few years ago. Again in 1901 at the same show Mr. Ainscough, we are informed, sold a cockerel for £100. Returning to brown-reds, we remember some eight or nine years ago that fourteen brown-red Game changed hands for the princely sum of £700, or £50 per bird. These, of course, are records and not every-day sales, still they go to show what really first-class exhibition specimens of Game fowl will fetch, and prove what the Yankee says—there is money in them.

Until the last year or two the brown-reds, like the brown-red Bantams, were in the hands of two or three, but now things are altered. Young fanciers and breeders have persevered until they have come to the front, so that at *the* Game show of the world—viz., Birmingham—we find old stagers behind breeders of only three or four years' experience. It is very fortunate that this is so, as it gives the breed a stimulus just at a time that it was in danger of losing ground owing to the deck-sweeping of one or two exhibitors.

Brown-reds are not, if properly understood, so very difficult to breed. In fact, quite recently we heard one party at a show, well known for the excellence of his strain, affirm that so true did they breed to the parents, that his great difficulty was, not so much to produce a winner, but to find in what way one bird excelled another; they are, as he said, like "peas in a pod". But it must be understood that this fancier knows very thoroughly the way he is working. I do not mean to say that any novice who has a mind to pick up any cheap specimens and mate them together irrespective of strain will produce a yard full of winners, or even a single one, but, rather, that when a strain has really been got together, first class as to quality, and the strain thoroughly under control in its owner's hands, it can be made, in the produce, to show a fine percentage of winners, second to no other variety in this respect, and such as to gratify the most expectant mind.

And now to describe the brown-red. In the Game points of size, style, shape and reach, they are Game birds all over. It is merely the colour that makes the bird a brown-red as distinguished from a black-red or a pile.

In colour the cock approaches the former in that he is a mixture, black and warm colour. If he be turned with his breast towards you he is easily recognised at a glance. It is a rich black, lightly but regularly laced round each feather from the throat down to the thighs, with pale gold lacing, and that I may not have to revert to this point again in the hen, I may here say that she is a rich black all over, save for two points, this same lacing on her breast and her neck hackle, which, like that of the cock, is a pale lemon, but striped narrowly with black. The breast feathers also have the shaft showing clearly.

To finish off the description of the cock I may say that his saddle is a rich but pale lemon to match the neck hackle, whilst the wing bows and back are a deeper but still very rich lemon colour. Unlike the black-red, he has no bays on wings, and probably on this account is a less taking bird with some than the black-red, which, taken all in all, has a larger proportion of rich colour, contrasting with the black, than the brown-reds have. Still the lemon top colour is very taking indeed. Formerly it was more or less of an orange, but that has long since been a thing of the past, though the superior brightness of present-day stock has doubtless been acquired at some loss of hardness of feather and shape, for it will invariably be noticed that, the lighter a bird runs in colour, the softer is the handling. So that the new lemon tinge is not (beautiful as it appears) an unqualified success. I contend that shape and type (which includes hardness of feather) should come before considerations of colour, otherwise we are getting more rapidly away from the original Game idea than ever.

With this brief description I may pass on to the mating, which is the most important part of the programme.

Again, two pens are necessary. For cockerel breeding we want a perfect exhibition cock or cockerel, a black-eyed mulberry-faced bird with bright top colour, plentifully bestowed, and the clearest breast lacing, supplied in no niggardly fashion. But at the same time he should have clear black shoulder butts, and herein lies the great difficulty. Quite 50 per cent. of the present-day exhibition brown-red cockerels fail in this respect: they are laced on the shoulder butts instead of being black, and this is a serious fault. Try and guard against this as far as you can, yet have the top colour as profuse and as rich a lemon as possible. Bear in mind that you will get your colour from the male bird, and the shape from the hens. To a cock or cockerel as I have described mate cockerel-breeding pullets. In the first place they must be typical Game shape, tall, square-shouldered, wing butts prominent and carried high, short in back, and narrow towards the stern, long in shank and sound in feet, hackle as pale a lemon right up to the crown of head as is possible to get. Never use dark-capped pullets for cockerel breeding, as it is labour in vain. If the pullets are laced on shoulder and back so much the better, so long as they are a pale lemon to top of head. About three pullets to a cock will be found most satisfactory for cock breeding, as you want size and reach.

For pullet breeding great discretion is required to select a cock of pullet-breeding strain. Go to some breeder who has figured most prominently in pullets at our leading shows, and buy a cockerel the same way bred as his winning pullets. As a rule pullet breeding cocks are a much darker top colour, more of a dark orange than lemon. See that he is specially good in breast lacing, even and distinctly laced from throat down to thighs, not blurred or indistinct. Shoulder butts and wing bar must be a sound black, no lacing whatever; this is highly important. He should be as shapely and as tall as you can get him, but, as I said before, it is essential that he should be of a pure pullet-breeding strain. To such a bird you must mate exhibition pullets, *i.e.*, pullets perfectly sound in colour. They must be tall and typical in shape, prominent shoulder butts, carried high, pure green-black body colour, free from shaftiness or lacing. In cock-breeding pullets the centre shaft of the feather shows lemon or golden; but in pullet breeding this is not so. The whole body and wings should be greenish-black, with lemon hackle, narrowly striped with black and as clear as possible towards the comb. Many pullets run quite coppery, or even darker, in cap. This should be guarded against, as it is a serious defect. Try and get them a pure lemon to crown of head; but, at the same time, take care that they do not show lacing on back for pullet breeding. In brown-reds, both cock and hen, the eye is an important feature. It must be as dark as dark can be. A light eye is a sure indication of a cross at some time or other.

If only one pen of birds can be mated up, and it is desired to breed both exhibition cockerels and pullets, then it will be a case of selecting two or three hens, such as I have described for cockerel breeding, and two or three exhibition hens, putting them all to an exhibition cock and looking for a fairly satisfactory result. But where it is possible to mate up two pens, do so by all means; still every working-man fancier cannot afford to do this.

Before leaving brown-reds I ought to advert to the introduction of birchen blood into many strains. It certainly did lighten matters up, and gave that bright lemon colour, but at the same time it greatly interfered with the dark gipsy face which is so essential to brown-reds. There is now no necessity to pursue such a course, as the desired tint has been secured long ago, and therefore there is really nothing to gain by resorting to birchen cross. If your stock is not bright enough, why then invest in a rich bright-coloured cockerel, and mate up as I have already stated. If you have to give your bottom dollar for it, it will pay you in the long run. There are plenty of bright lemon cockerels ready to hand. In these busy times of keen competition there is no time for retrogression to antique ways, and so with this piece of advice I will pass on to piles.

### Piles.

We now come to the most beautiful of all varieties of Game, but, at the same time, to the one admittedly the most difficult to breed.

Beautiful as are pile Game, probably by reason of the sharper colour contrast which exists in them than in the other varieties, they have just this one great drawback, good specimens are not so easily produced, notwithstanding the most careful mating. A much larger flock of youngsters has to be reared annually than in the other varieties before a "flyer" can be produced, and when he is produced he does not command as high a price as a first-class black-red or brown-red, although both the latter are easier to produce and easier to keep in show condition than the pile. Still, notwithstanding these drawbacks, this beautiful variety finds many admirers, who do not seem to mind the extra trouble of preparing for the show pen, and which is no small task unless kept under most favourable circumstances.

Like all birds with much white in their plumage, they require a bath to get them into proper show condition; at the same time the cleaner they can be kept, so as to relieve them from the feather-softening influence of soap and water, the better.

It is possible, where the fancier has a good clean grass run away from the smoke of large chimneys, with due care to run a bird at many shows held about the same time with very little washing, but it must be protected from the weather, and be provided with the cleanest of runs and houses, otherwise the tub will be a *sine quâ non*, for in these days of severe competition no exhibitor can afford to throw a single point away in the matter of exhibiting his birds. Therefore, where the young fancier or beginner cannot command a run such as I have described, I would not recommend him to take up pile Game. If he is near the smoke of a manufacturing town let him take up black or brown-reds if he is desirous of becoming successful in the exhibition pen.

Piles in length of limb probably surpass other varieties in the best exhibition specimens, otherwise in all points of shape, style, size and reach, the description given for Game birds in general holds good with them too.

A good pile, therefore, must be of the tallest, short and hard in feather, wedge shape in body, high and prominent in shoulder, broad chested, cleanly cut away behind, with short wings clasped tightly to the body, a neat whip tail carried slightly above the level of the body and as fine in sickle as possible. Head long, lean and snaky with a bold alert red eye. To easily describe the colour of the exhibitor's pile cock I must refer my readers back to the chapter on black-reds, for where a first-class exhibition black-red is *black*, the pile cock should be *white*, the top colour precisely the same in both. I have seen piles with a top colour all through one dark brick-red colour winning at the Palace, but this is wrong. A black-red of this colour would not even be mentioned in the prize-list however good he might be in other points; then why should a pile?

The two great failings in piles are the difficulty in obtaining a perfectly sound white breast free from lacing, ticking or smokiness, together with a rich top colour and dark chestnut bays, or wing ends. Quite 50 per cent. of piles fail in these two points, both of which are highly important in exhibition specimens.

A cockerel with a laced or smoky breast has very little chance of success in these days of keen competition, whilst weak bays are almost as great a drawback. The leg colour, too, is a very important point with some judges. This should be a rich orange yellow, the richer the better. Soil has a great deal to do with the leg colour. Where the land is a heavy clay, the leg colour will always be good, but where the soil is limestone and dry the yellow will always be much paler.

The pile hen should be a clear white all over save her breast and neck hackle. The breast may be described as a good warm-tinted chestnut, as deep in colour as possible, providing it does not affect the wing colour. It is one of the greatest difficulties to breed a pile pullet perfectly clear on wing, free from any tinge of red or creaminess, and at the same time having a rich, deep salmon breast. Where the body colour is specially clear, the breast will invariably be found too pale, and where the latter is very deep, the wing will be found to be tinged with red; still, where the body colour is absolutely clear, some allowance can be made for a little weakness in breast colour. The neck hackle should be a pale lemon.

Now for the breeding pens. A double one is usually required. It is more or less guesswork going on the one pen system. Necessity may drive a breeder to it, but I always advocate the dual system whenever it can be practised.

We will take the cock-breeding pen first. In choosing your male bird let him be an ideal exhibition specimen. Personally I much prefer the bright kind. See that his colour is sound, not only his top colour, but his breast a clear white, and his wing ends a rich chestnut right to the end of the feather. If the bays are washed out or weak, don't have him at any price—for cockerel breeding he would be dear as a gift. Remember always in your mating up that whatever faults your male bird may have in colour will crop up manifold in the greater part of the produce. Therefore be most exacting in your colour requirements, especially in the cock bird. If he is good in this respect, you can safely use him, providing he has the necessary length of limb and body shape. In selecting his mates let them excel in *shape* and reach, and the more stylish and limby the better. They must be sound deep salmon in breast colour and well rosed with red on wings. You can't have too much colour on wing; such pullets, although useless as exhibition specimens, are invaluable as cockerel breeders. For cock breeding it is best where size and reach are required to use not more than four pullets to one cock or cockerel; three would be better. From such a pen you may reasonably expect some first-class exhibition cockerels. The pullets from this pen will also be very useful the following season for cockerel breeding, mated back to their father, and the best cockerel may be used with the hens. In this manner you are building up a reliable strain of your own, which is the secret of success in breeding high-class specimens, and is a very important requirement.

Another way to breed exhibition pile cockerels is to procure a sound coloured exhibition black-red cock and mate him to two or three pale-breasted pile pullets. In choosing a cock the most essential requirements are a sound black breast and sound, unbroken, deep chestnut bays, combined with a rich bright top colour. Some of the very best pile cocks that have ever graced the show pen have been bred in this manner. The only drawback is that the majority, if not all, of the pullets will come willow-legged. Years ago a good coloured pile with willow legs would find a place in the prize-lists, but those days are gone. Willow-legged specimens, no matter how good in other points, are discarded in the show pen of to-day.

It will invariably be found that these willow-legged pullets are better in shape, harder in feather, and sounder in colour than the yellow-legged ones. For this reason, I have found them very useful for pullet breeding when mated to a reliable pullet-breeding pile cock, especially so when the leg colour has shown a tendency to go pale in the piles the previous season. Therefore, although useless as show birds, they must not be treated as "wasters".

For pullet breeding a darker top-coloured bird is desirable—one of a red-brickish colour all through on top; but he, too, must be sound in his wing ends. It is best to procure one from a strain that has produced the best pullets at the principal shows. Pullet-breeding cocks are, of course, of little use for exhibition, consequently they can be bought much cheaper. Mate him to pullets or hens, from four to seven in number, perfectly clear on wing, and possessing good salmon-

**BROWN-RED GAME PULLET.**

coloured breasts. See that they are tall and excel in shape; the latter point is most important. The following season mate the best pullets back to their father and the hens back to the best cockerel.

If the space at your command will only allow of one pen being used, then select as highly a coloured cockerel as you can find that has been bred from a pullet-breeding pen, and mate him to a couple of rosy-winged pullets as described in cockerel breeding, and two clear-winged, good-breasted hens. You will then breed a few exhibition specimens of both sexes.

Now a little advice in mating up your birds:— Never use delicate birds or unhealthy specimens. The cock should never be over three or four years old, and never less than twelve months. From one to three years is the best, if you desire fertile eggs and vigorous chicks. Bear in mind that unhealthy stock birds will breed weakly chicks that will be an eyesore to you during the little time allotted to them in this world, and the sooner they depart the better for every one concerned. When the colour in your pile cocks appears to be getting paler and washed out as it were from breeding year after year from piles on both sides, the remedy will be found by introducing a good coloured black-red cock into your next season's cock-breeding pen, instead of using a pile cock. This cross also will always improve the leg colour when it has become pale through in-breeding.

In-breeding is all right when judiciously performed, but there is a happy medium beyond which if you proceed it will result in disaster, both from a colour point of view, and more especially in the health of your produce.

### Duckwings.

Duckwings will now claim our attention. In my own estimation there is no more handsome bird in the whole poultry world than a sound-coloured golden duckwing cock.

Although not so difficult to breed to colour as the pile, yet he hardly secures the same favour as the latter in the Game Fancy. Still, a really first-rate duckwing can always hold his own in the show pen even when pitted against the more popular black-red, on account of the rich contrast in his colours. There are golden duckwings and silvers, the latter being serviceable only as stock birds in this country, but in America, Belgium and other countries both golds and silvers are catered for with separate classes in the show pen.

But here in England a silver duckwing has no earthly chance of success as an exhibition bird; still, as I shall point out later on, he is useful as a pullet breeder when properly mated. The day may come when silver duckwing Game, like the silver duckwing Leghorn, will find a place in the exhibition world.

I will now go on to describe the golden duckwing cock. In shape, size and general characteristics he must be the same as described for black-reds. He should have a red face and eye, although there is a great tendency for duckwings to run dark in eye. This, of course, is a fault, and should be guarded against when mating up your breeding pens. The neck hackle and saddle hackle should be a silvery white, as free from striping as possible. Saddle and wing bow a rich, deep orange, shaded with maroon, the richer and deeper the colour the better. The breast, wing butts, wing bar, and thighs should be a sound black, free from any lacing or ticking. The wing ends from the black wing bar to the end of the wing should be clear white, free from any chocolate colouring on the outer edge. It is in this point that quite a large percentage of duckwing cockerels fail, and the failure is somewhat difficult to overcome, and requires most careful attention when mating up. The tail of the duckwing should be identically the same as in the black-red.

The duckwing pullet, with the exception of colour, should resemble the black-red. The colour should be a nice light shade of steel grey, one even soft shade throughout, as finely pencilled as possible, although pullets bred from a cock-breeding pen will invariably fail in this respect, the coarse marking and blotchiness appearing on the top flight feathers of the wing, which also is a very great defect in the exhibition pullet. The legs should be willow, and as round and fine in bone as possible, flat shins being objectionable in all Game fowl. The breast colour should be the same as in the black-red, a medium shade of broken salmon. It is almost an impossibility to secure the deep salmon breast and the nice soft, even shade of steel grey body colour combined. Exhibition duckwing pullets have a great tendency to run pale in breast, and where the wing colour is particularly sound and free from any rust or coarse markings, some allowance should be made for a little paleness of breast colour. When I say a little pale I do not mean a regular washed-out breast colour, which we sometimes see heading the prize-list. This is wrong, as it is quite an easy matter to get a sound wing colour and a very pale breast, but quite another matter to combine a *fairly good* breast and sound wing colour.

# DUCKWINGS.

And now to mating. It is much the best to have two pens—in fact, almost an absolute necessity.

For breeding the best-coloured duckwing cockerels, it is best to use a sound-coloured, black-red cock, one perfect in his black, and possessing a sound, rich, deep bay to end of wing. Of course, it is highly essential that he should be a typical Game bird in shape and reach as described in former chapters, with a straight breast-bone and sound feet.

To this bird I should mate two or three tall, broad-shouldered, duckwing pullets, as tall as possible, and preferring those clearest in neck hackles. If coarse in markings, and inclined to be a bit rusty on the wing, all the better for producing plenty of colour in your cockerels; but, above all, they must be tall and excel in shape.

From this pen you should breed some good-coloured, reachy cockerels, possessing all the requirements of a first-class exhibition bird. The pullets, however, from this pen would be mostly black-reds, and the few duckwing pullets that were produced would be faulty in colour from an exhibition point of view, but useful as cock breeders another year if mated to a rich, deep-coloured duckwing cockerel.

In pullet breeding I should go on quite contrary lines by using sound-coloured, shapely black-red pullets, mated to a silver duckwing cockerel. Not only would you breed good duckwing pullets in this manner, but in all probability you would also get a fair percentage of soft, even-coloured, black-red pullets, good enough to hold their own in the show pen; especially so if the black-red pullets you used in the breeding pen were pure pullet-bred black-reds. The most important point in this mating is to use only a silver duckwing cockerel that is pure duckwing bred on *both sides*—*i.e.*, bred from a duckwing cock out of a duckwing hen. Duckwing pullets can also be produced by using exhibition duckwing hens mated to a duckwing cock of the medium shade of colour, but the hens must be sound in colour and free from shaftiness—a very common fault in duckwings, and a fault which should by no means be overlooked in mating up your breeding pens.

The black-red pullets bred from a duckwing cross should never be used for breeding in the pure black-red breeding pens; from a colour point of view their produce would be worthless. Therefore, if the young fancier is breeding both duckwings and black-reds, the black-reds bred from duckwings should be marked as soon as hatched, and afterwards rung, to distinguish them from the pure-bred black-reds.

## Birchens.

The birchens are closely related to the brown-reds, and were originally produced by crossing the silver duckwing cock with brown-red hens with a view of improving the colour of the brown-red cocks from a deep orange colour to the present day pale lemon. Birchens only differ from the brown-reds in top colour and lacing. In the former it should be a silvery white where the brown-red is lemon; in all other points they are precisely the same. Although a very handsome variety, the birchen finds few admirers, and very rarely are classes provided for them, with the exception of Birmingham, the Game show of the world.

Competing against piles and duckwings, they stand little chance of success in the prize list, and for this reason they have failed to find favour. Birchens breed more true to colour than either piles or duckwings, and good specimens of both sexes can be produced from one pen. The most difficult points to secure are a sound silvery white top colour in cocks, with a dark eye and a dark mulberry face combined. From a financial point of view I would not recommend any one to take up birchens, as prices for really first-class birds are by no means high.

## Whites and Blacks.

White Game are what their name denotes. Plumage pure white throughout both in cocks and hens, with rich yellow legs and beaks, and red eyes; in other points—viz., style, reach, shape and carriage—they must resemble first-class exhibition black-reds.

At the present time (1902) they are few and far between.

During the past few years I have visited hundreds of shows and have not come across more than a dozen specimens at the most, and these were at Birmingham, the only show where classes are provided for whites and blacks.

Wales was at one time famous for its white Game, and I believe there are still a few bred somewhere in South Wales, but they are very rarely exhibited. Handsome as they are, they find few lovers, doubtless on account of the difficulty of keeping them clean, and are only suitable where they can have a good grass run away from the smoke of large towns. Where kept under favourable circumstances they are a very pretty variety, and look exceedingly nice on a good grass run, and are by no means difficult to breed or rear, and, moreover, one pen only is required to produce good specimens of both sexes.

Blacks are still more scarce than whites, and are almost extinct. The colour should be a lustrous black throughout; legs and feet, dark bronze or black, the blacker the better; beak to match the legs; eyes dark brown, as dark as possible; face, comb and lobes, red. The only redeeming feature there seems to be about blacks is that they do not show the dirt, and therefore are suitable for confined runs in towns, where piles, birchens, or whites could not be kept; but apart from this, there is nothing to recommend them to the young fancier, as the small demand for this breed would not pay him for his trouble and outlay.

# THE STANDARD OF PERFECTION FOR MODERN GAME.

The following is the Standard of Perfection reproduced by kind permission of The Poultry Club:—

### GENERAL CHARACTERISTICS OF COCK.

*Head and Neck.*—Head, long and snaky, narrow across the eyes; beak, strong at base, gracefully curved; eye, prominent, keen in expression; comb, earlobe and wattles, usually cut off; face, smooth skin without coarseness; neck, long and slightly arched, fine at setting on of head.

*Body.*—Short, wide in front, well tapered to stern; breast, broad; back, flat and shaped like a smoothing iron; wings, strong and powerful, short, well tucked up, shoulders prominent and carried well up.

*Tail.*—Short and fine, closely whipped together and carried slightly above level of body; sickles, fine and well pointed.

*Legs and Feet.*—Thighs, strong and muscular; shanks, long and nicely rounded; toes, long and straight.

*General Shape and Carriage.*—Upstanding and active. General appearance bold, fearless and smart.

*Size and Weight.*—From 7 lb. to 9 lb.

*Plumage.*—Short, hard and bright.

### GENERAL CHARACTERISTICS OF HEN.

*Head and Neck.*—Generally corresponding with cock; comb, very small and erect, nicely serrated; earlobe and wattles, small and fine texture.

*Body.*—Back, flat; rest of body corresponding with cock.

*Tail.*—Short and fine, closely whipped together and carried slightly above level of body.

*Legs and Feet.*—As in the cock.

*General Shape and Carriage.*—As in the cock.

*Size and Weight.*—From 5 lb. to 7 lb.

*Plumage.*—Short, hard and bright.

### Colour of Black-breasted Red Game.

*In both Sexes.*—Beak, dark green horn; eye, bright red; comb, face and wattles, bright red; earlobe, bright red; legs, willow green.

*In the Cock.*—Head, orange red; hackle, light orange and free from black stripe; breast and thighs, greenish-black; back and saddle, rich crimson; wing-bow, orange; shoulders, black; wing-bars, green-black; secondaries, rich bay on the outer edge of feathers, on the inner edge and tips of bay black, only the rich bay showing when the wing is closed; primaries, black; tail, sickle feathers and tail coverts, green-black.

*In the Hen.*—Head, gold; hackle, gold slightly striped with black, running to clear gold on the top of the head; breast and thighs, breast a rich salmon, running to ashy colour on thighs. Rest of body, a light partridge brown with very small markings, and a slight golden tinge pervading the whole, which should be even throughout, free from any ruddiness whatever, with no trace of pencilling on the flight feathers; tail, black except the top feathers, which should match body colour.

### Colour of Red Wheaten Game Hen.

Beak, greenish horn colour; eyes, ruby red; comb, lobes, face and wattles, red; legs, willow; plumage, head and hackle, golden or lemon, very slightly striped with black; breast and thighs, fawn or cream, diminishing to pale buff on thighs; body colour, pale cinnamon or wheaten; secondary flight feathers, pale cinnamon or wheaten; tail, black, except top feathers, which match body colour.

### Colour of Silver Wheaten Game Hen.

Beak, light horn colour; eyes, ruby red; comb, lobes, face and wattles, red; legs, willow; plumage, head and neck hackle, silvery white, very slightly striped with black; breast and thighs, pale fawn, diminishing to light buff on thighs; body colour, very pale cinnamon; secondary flight feathers, very pale cinnamon; tail, black, except top feathers, which match body colour.

# MODERN GAME.

### Colour of Brown-breasted Red-Game.

*In both Sexes.*—Beak, very dark horn, black preferred; face, legs and tail, black; eye, jet black.

*In the Cock.*—Head, extra rich bright lemon; hackle, bright lemon, the centre of the feathers striped with glossy green-black colour, not brown; breast and thighs, brilliant glossy green-black, the breast feathers edged with round pale lemon lacing as low as the top of the thighs. Back and saddle, pure bright lemon; shoulders and wing-bows, points of shoulders very glossy green-black, free from ticks or lacing. Back of shoulder and wing-bow pure bright lemon to match back and saddle; wing-bars, rich glossy green-black; rest of body, very bright glossy green-black.

*Note.*—There should only be two colours in brown-red Game, *viz.*, lemon and black. The lemon colour in the cock should be very rich and bright; in the hen it should be a light lemon. The black in both sexes should be a very rich bright green-black called a beetle green.

*In the Hen.*—Head and neck hackle, pure bright lemon right to top of head, the lower feathers striped with glossy greenish-black colour; breast, laced same as cock well down to the thighs with light lemon-coloured lacing; rest of body and tail, rich beetle green-black, free from ticks on shoulders or lacing on back.

### Colour of Pile Game.

*In both Sexes.*—Beak, yellow; eye, bright cherry red; face, red; legs, rich orange yellow.

*In the Cock.*—Head, bright orange; neck hackle, bright orange colour (dark washy hackles are to be avoided); breast and thighs, pure white; back and saddle, rich maroon; saddle hackles, bright orange; shoulder butts, pure white; wing-bows, rich maroon; wing-bars, pure white, free from splashes; secondaries, dark chestnut on the outer edge of feathers, on the inner edge and tips of bay white, the dark chestnut only showing when the wing is closed; primaries, pure white; tail, sickle feathers and tail coverts, white.

*In the Hen.*—Neck hackle, white tinged with golden colour; breast, salmon colour; rest of body, pure white.

### Colour of Golden Duckwing Game.

*In both Sexes.*—Beak, dark horn colour; eye, ruby red; face, red; legs, willow.

*In the Cock.*—Head, creamy white; hackle, creamy white, free from striping; breast and thighs, blue-black; back and saddle, pale orange on rich yellow; wing-bows, pale orange on rich yellow; wing-bars, black with blue sheen; secondaries, pure white on the outer edge of feathers, on the inner edge and tips of bay black, the pure white alone showing when the wing is closed; primaries, black; tail, sickle feathers and tail coverts, blue black.

*In the Hen.*—Head, silvery white; hackle, silvery white, finely streaked with black; breast and thighs, salmon colour, diminishing to ashy grey on thighs; rest of body, French or steel grey, very slightly pencilled with black and even throughout; tail, black except top feathers, which should match body colour.

### Colour of Silver Duckwing Game.

*In both Sexes.*—Beak, horn colour; eye, ruby red; face, red; legs, willow.

*In the Cock.*—Head, silvery white; breast and thighs, lustrous blue-black; hackle, silvery white; back and saddle, silvery white; shoulder coverts and wing-bows, silvery white; wing-bars, steel blue; secondaries, pure white on the outer edge of feathers, on the inner edge and tips of bay black, the pure white alone showing when the wing is closed; primaries, black; tail, sickle feathers and tail coverts, blue-black.

*In the Hen.*—Head, silvery white; hackle, silvery white, with very narrow black stripes; breast and thighs, breast a pale salmon, diminishing to a pale ashy grey on thighs; rest of body, light French grey, with nearly invisible black pencilling; tail, black, except top feathers, which should match body colour.

### Colour of Birchen Game.

*In both Sexes.*—Beak, dark horn; eye, black; comb, face and wattles, dark purple; legs, black.

*In the Cock.*—Plumage, head, silvery white; hackle, silvery white, with narrow black striping; breast and thighs—breast a rich black with a narrow silvery margin round each feather, giving a beautiful laced appearance gradually diminishing to perfect black thighs; back and saddle, silvery white; shoulder coverts and wing-bows, silvery white; wing-bars, glossy black; secondaries, glossy black; tail, sickle feathers and tail coverts, black.

*In the Hen.*—Plumage, head and hackle, silvery white with very narrow black stripes; breast and thighs—breast black, very delicately laced with white, diminishing to perfect black thighs: rest of body and tail, lustrous black.

Value of points in Game:—

| Defects. | Deduct up to |
|---|---|
| Defects in head | 5 |
| ,, neck | 5 |
| ,, eye | 10 |
| ,, tail | 10 |
| ,, legs and feet | 10 |
| ,, colour | 20 |
| Want of style and shape | 30 |
| Want of condition and shortness of feather | 10 |
| A perfect bird to count | 100 |

Serious defects for which a bird should be passed:—Eyes other than standard; crooked breast-bone; twisted toes or duck feet; wry tail; crooked back; flat shins.

## HATCHING AND REARING.

If I were asked which breed of large fowl were the most difficult to rear, I should unhesitatingly reply Modern Game. Doubtless many of my readers will remark, "Why is this?" The chief reason is not far to seek, *viz.*, "In-breeding".

The present-day Modern Game fowl has been so inbred to obtain the desired colour and fine bone, and bred so very tall, that the constitution and stamina of the bird has been so enfeebled as to make the chicks most difficult to rear, except where every care and attention can be given to them by an expert. Chicks bred from birds kept especially for stock purposes are much easier to rear than those bred from exhibition birds that have been doing the rounds of shows the previous season. Therefore I always advise young fanciers to breed from stock that are healthy and vigorous, and not from worn-out exhibition specimens which are worthless as breeders.

Do not try and hatch your Game chicks in incubators and then rear them in an artificial rearer. You may rear Rocks, Orpingtons and the like in this manner, but pure-bred exhibition Game, *never*. They require a natural mother, and a good one to boot—nothing beats the Wyandotte-Silkie cross. Mate two or three silver or gold 'Dotte hens to a Silkie cock, and the pullets from this cross can be relied upon at all times, and will cover nine or ten Game eggs nicely; and, moreover, will be broody by the time your first batch of eggs are ready to put down—instead of having to tramp the country round in search of a "clocking" hen, which, when you have found her and taken her home, very probably will prefer to stand on the eggs rather than sit.

Having got your eggs and your broody hen ready, proceed to make the nest. Nothing beats the old-time orange box, which can be bought for 2d. or 3d. Place the box on its side with a strip of board about 3 or 4 inches deep at the bottom to hold the nest material, and to keep the eggs from rolling out. Next put in a good supply of sand or fine soil, make the soil hollow in the centre, and heaped round the sides of the box, then put in plenty of nice soft hay, making it the shape of a basin.

Before putting the hen on—which is best done at eventime, when it is getting dusk—examine the hen for lice; if she has any, give her a good dusting with insect powder; also sprinkle some on the nest before putting in the eggs. Take the hen off each evening, and give her a feed of maize, and water to drink. Do not allow her to go on until the wants of nature have been attended to, or she will foul the nest. At the end of the sixth day, examine the eggs by candlelight, as by that time the clear or unfertile eggs can easily be detected simply by holding the egg between the thumb of each hand, and turning the egg slowly round, when the fertile eggs will show a dark shadow in the centre, whilst unfertiles will be quite clear. These can be taken out to make more room for the fertile ones.

When possible it is best to set two hens at the same time, then if there are many clear eggs, all the fertile ones can be given to one hen, and a fresh lot of eggs to the other; by doing this, you will ensure good broods. If any of the eggs remain unhatched at the end of the 21st day, it is best to get a basin of luke-warm water and put the eggs into the water, and if the chick is alive the egg will bob about, generally within the space of a minute or two at the most. If they remain stationary, you can be sure the chick is dead.

In very dry hot weather it is a good plan to take out the eggs and damp the nest well with milk and warm water a few days before hatching, repeating every alternate day.

Just before the hen is due to hatch, examine her again carefully for lice, and sprinkle her well with the insect powder. More deaths are caused through lice on chicks than all the diseases put together, therefore you cannot be too careful in this matter. Cleanliness is the certain road to success in chicken rearing, so don't you forget it. Always provide a good dust bath for your sitting hen, composed of dry ashes with an addition of a little sulphur.

After the chicks are hatched, take the hen and chicks and place them in a coop with a boarded floor covered with nice clean dry sand; let her remain perfectly quiet for a few hours, as chicks do not require any food for at least twelve hours after hatching.

### Feeding Game Chickens.

The first feed should consist of hard boiled egg chopped fine, and mixed with stale bread crumbs. This should be given to them every two hours the first day.

The next day or two feed them on Game Meal mixed with boiling water and allowed to cool. Give them this for the first week, varying the feed once or twice a day with bread soaked in sweet milk, and, for a change, once a day rice boiled for fifteen minutes, and after it has had the water taken from it and allowed to cool, rub it through fine oatmeal until dry. These two last feeds will greatly assist in preventing diarrhœa. After the first week the feeding can be reduced to every two-and-a-half hours. To the meal add half the same quantity of Indian meal, and a handful of meal greaves or crissel, and scald both together; stir well together, and allow to stand a little while, and then add a handful or two of oatmeal and fine sharps, or what is termed thirds flour: this makes a capital staple soft food for Game chicks after the first week until they are matured. At the end of a month the feeding can be regulated to four times daily; the feed at night being wheat, which must be sound and sweet.

And now for the water question. Where the chicks have free range on grass runs, do not give them any water at all; they will thrive all the better without it. The majority of our big successful breeders never give their chicks water, consequently diarrhœa is almost unknown. Give them plenty of water and allow it to be exposed to the sun in warm weather, then your troubles begin and your chicks will die off by the score. Never give them water, and they will never require it. Of course the hen will require water; this can be done by fastening a small tin inside the coop about eight inches from the ground. When you give your chicks their first feed in the morning do not forget to feed the hen at the same time; give her a good feed of maize then, it will satisfy her for the day, otherwise she will eat up the food intended for the chicks.

When the hen leaves the chicks, which is generally at the age of from two to three months, the cockerels should be separated from the pullets, and placed in a separate run; by doing this the chicks will thrive much better. Any deformed or mismarked ones should be killed at this age to make more room for the good ones.

## DISEASES.

WE now enter upon the concluding chapter of our work by a brief reference to the diseases to which Game fowl are liable.

Old English Game, being of a more robust constitution than Modern Game, rarely suffer much sickness. They are altogether stronger, and more capable of withstanding the cold and damp than Modern Game, the stamina of which, through in-breeding, has been greatly weakened.

The most common of all ailments to which exhibition Game are subject is COLD or CATARRH. The chief symptoms are a watery discharge from the eyes and nostrils, accompanied by feverishness, and if neglected in its early stages will develop into roup. As soon as the symptoms have been noticed, the patient should be isolated at once, and penned in a well-ventilated house free from draughts. The eyes and nostrils should be bathed night and morning with warm water to which has been added a few drops of permanganate of potash or Condy's Fluid. Should the nostrils be full of sticky mucus, this should be pressed out with the forefinger and thumb; afterwards the mouth should be washed out as before, using a piece of linen rag, which should always be burned after each operation to prevent the spread of the disease. Give the bird a teaspoonful of salad oil each morning half-an-hour before feeding, and again at bedtime, but never when the crop is full of food.

The morning food should consist of bread soaked in cold water; then squeeze out the water and add a little warm milk. At night give good, sound, dry English wheat, but feed sparingly. The drinking water should have a small piece of sulphate of iron about the size of a small bean to a pint of water, but only allow the bird to drink twice a day after food. Keep the bird supplied with both green food and grit, with a little lean meat, cut fine, every alternate day.

DIPHTHERITIC ROUP OR DIPHTHERIA.—This loathsome disease is the most contagious to which the feathered tribe is liable, as well as the most fatal. I have known it

# DISEASES.

to go through the entire flock, causing a loss in some cases of two or three hundred pounds. The symptoms are a whitish film or membrane in the inside of the mouth, which gradually thickens until it becomes of a cheesy nature, and spreads down to the windpipe. When this stage is reached, the bird has a catchy cough, also arching and twisting its neck in a peculiar manner, and unable to swallow its corn. If the growth is not immediately removed the bird is suffocated. Besides the growth in the mouth, wart-like spots will appear on the face and round the eyelids to such an extent that the bird goes completely blind. These ulcers or sores like warts may appear on the face and at the same time the mouth be perfectly free.

In the first place, the bird should be isolated immediately the disease is noticed, and the house thoroughly disinfected and lime-washed, adding a little carbolic oil to the lime-wash. The drinking vessels should be scalded with boiling water, adding a little Jeyes' Fluid to the water, and allow them to remain in the water for some minutes.

The mouth and spots on the face and comb should be painted with the following lotion, which can be obtained at any chemist's:—

| | |
|---|---|
| Carbolic acid | 1 drachm |
| Sulphurous acid solution | 3 drachms |
| Tincture perchloride of iron | 4 drachms |
| Glycerine | 4 drachms |

using a camel-hair brush.

The cheesy matter should first of all be removed, using a piece of sharp wood and burning the wood as soon as you have finished. Then paint the places carefully morning and night. It is not necessary to remove the growth every day —only when it appears ripe.

In addition to this treatment make a few pills of lard and flour of sulphur—equal parts—and give one the size of a small marble each morning for a week.

The hands should be well washed and disinfected after each operation, as the disease is highly contagious, and the clothes should be changed before going amongst the other birds which are not affected. I have known people carry the disease in their clothes for miles, even though they had not handled the birds.

If the first bird affected is not a very valuable one, the best plan would be to kill it and burn the body. By doing this you would in all probability prevent the spread of the disease, if the place be at once thoroughly disinfected. I have found Krekodyne vinegar added to the drinking water a capital preventive of the spread of this disease.

BRONCHITIS.—Symptoms: quick breathing and coughing, with frothy discharge from the mouth and nostrils. It generally comes on very suddenly if the bird has been exposed to a draught or severe change of the weather. The patient should be kept in a warm room, and allowed to inhale steam from boiling water into which has been added a few drops of terebene. Repeat the treatment every two hours until the breathing is easier. Feed the bird on warm bread and milk.

LIVER DISEASE is one of the most prevalent of diseases to which Game fowl, through in-breeding, are liable. The symptoms are sluggishness, little or no appetite. When once this disease has got a firm hold it is almost impossible to cure it, but by judicious feeding, and an occasional dose of salts or fluid magnesia, the disease may be kept in check. Green food is a necessity, and the morning feed should be bread soaked in water, or a little oatmeal porridge. Maize should on no account be given. In severe cases the bird may go lame.

INFLAMMATION OF THE LIVER is more acute, and it is rarely the patient recovers. The causes may be through wrong feeding or the result of a chill. There is much more pain, and the bird will appear to be unable to move about; very often there is lameness. A steam bath, by holding the patient over boiling water every two hours, will relieve the pain and inflammation; give ten drops of chlorodyne every four hours for three or four days. After the bird has got a turn for the better, feed on bread and milk in the morning, a little lean meat chopped fine with boiled cabbage at noon, and sound corn at night, with a little iron in the water as a tonic.

PNEUMONIA, OR INFLAMMATION OF THE LUNGS.—The symptoms are very similar to bronchitis: difficult breathing, panting and coughing. The bird must be kept very warm, and fed on warm bread and milk only; a little liniment or turpentine rubbed well between the shoulders. Give the bird three drops of ipecacuanha wine in a teaspoonful of water every hour until relieved. If the patient cannot eat, beat up a raw egg and add a little brandy; give this by using a small syringe, which can be purchased from any chemist for a shilling.

LEG WEAKNESS generally affects the Modern Game cockerels more than pullets, and is said to be caused by too much uric acid in the blood; that is to say—is of the nature of rheumatism. I am inclined to think it is simply the bird overgrowing its strength, and, being long in limb, the weight of the body causes the legs to give way. One of the finest remedies I know of is a teaspoonful of Easton's Syrup in the bird's soft food in the morning, and the same quantity in half a pint of drinking water. Another remedy is to make thirty pills from the following, and give one two or three times a day: Strychnine, 1 grain; citrate of iron, 1 drachm; phosphate of lime, 1 drachm; quinine disulphate, 15 grains. A useful preventive is a little bone meal in the soft food.

BUMBLE-FOOT.—This particularly attacks Dorkings and heavy breeds in general, but is sometimes found also in Game. The causes may be cuts from glass or pricks from thorns, when the formation is purely a common abscess. Corns will develop into abscesses. This is probably true bumble-foot. Flying down from high perches on to the hard ground, or any other action by which the foot is suddenly strained, may tear the corn until it festers, and must then be removed by cutting it away in one piece. If it is a soft abscess, make an incision and squeeze out the matter. Wash well with hot water, using plenty of soap; afterwards pour a few drops of Condy's Fluid into the incision, then bind up the foot with a piece of rag which has been smeared with carbolised vaseline.

SCALY LEG.—Purely and simply a parasite growing under the scales of the leg. Wash the legs thoroughly in hot water, as hot as you can bear your hand in; let the legs remain in the water until well soaked; then take an ordinary nail-brush and soap, and scrub well under the scales. If the disease is very bad it is best to apply pure paraffin to the legs a day or two before you wash them. After you have given them a good washing, dry thoroughly, and apply a little ointment made of equal parts of sulphur and vaseline. Repeat the dressing after an interval of three days. Scaly leg is by no means a formidable thing. It is very often contracted from the sitting hen when the chicks are very young. It should, however, be attended to as soon as discovered, otherwise it eats into the leg in time, and, when cleared off, leaves it pretty much as rust that has been long neglected would leave a piece of old iron.

CROP-BOUND is caused by the bird picking up portions of grass, hay, straw and similar material, which do not digest, but form into a hard ball in the crop. If taken in time, before it sets very firmly, the crop may be partially filled with warm water and the offensive matter kneaded to a pasty state. If the fowl be then inverted, the whole contents of the crop may be ejected by squeezing it out through the mouth. If this method fails, make an incision from the outside. Choose a place as free from veins as possible, then make a cut 1½ inch long, cutting the top skin and the under-skin together; by inserting the forefinger, the contents can easily be removed, taking care to get it thoroughly empty. The edges of the wound should then be washed with warm water to which a little Condy's Fluid has been added; then take a fine needle and some white silk thread and stitch the edges of the under-skin first, then the outer skin. Do not give the bird any water for a day or so; feed only on bread soaked in milk for a few days, until the wound has healed.

www.ingramcontent.com/pod-product-compliance
Lightning Source LLC
Chambersburg PA
CBHW062227220526
45471CB00009B/3382